THE BIG iDEA

HAWKING

AND

BLACK HOLES

PAUL STRATHERN

ANCHOR BOOKS

DOUBLEDAY

New York   London   Toronto   Sydney   Auckland

AN ANCHOR BOOK
PUBLISHED BY DOUBLEDAY
a division of Bantam Doubleday Dell Publishing Group, Inc.
1540 Broadway, New York, New York 10036

ANCHOR BOOKS, DOUBLEDAY, and the portrayal of an anchor are
trademarks of Doubleday, a division of Bantam Doubleday Dell
Publishing Group, Inc.

Hawking and Black Holes was originally published in the United
Kingdom by Arrow Books, a division of Random House U.K. Ltd. The
Anchor Books edition is published by arrangement with Arrow Books.

Library of Congress Cataloging-in-Publication Data

Strathern, Paul, 1940–
Hawking and black holes/Paul Strathern.
p.   cm.—(Big idea)
Includes bibliographical references.
1. Hawking, S. W. (Stephen W.)—Biography.   2. Black holes
(Astronomy)   3. Physicists—Great Britain—Biography.   I. Title.
II. Series.
QC16.H33S77   1998
503′.092—dc21
[B]   97-52136
CIP

ISBN 0-385-49242-1
Copyright © 1997 by Paul Strathern
All Rights Reserved
Printed in the United States of America
First Anchor Books Edition: August 1998

1   3   5   7   9   10   8   6   4   2

Contents

Introduction

STEPHEN HAWKING has been likened to Dr. Strangelove, the bogeyman of Kubrick's classic film. And there is more than the obvious passing resemblance. Hawking is of course no angst-ridden Nazi. Yet those who have worked with him speak of a similar intensity of suppressed intellectual energy. Dr. Strangelove was a parody of naked willpower—but of a complex, farseeing, largely cerebral variety. At the same time Dr. Strangelove was also utterly human, possessed of strong feelings and human foibles—which his crippling disability did nothing to lessen. Hawking has always insisted that he too should be viewed as a normal human being, and his actions have fully justified this view.

In the film we never see Dr. Strangelove's office. Had a location been required, Dr. Hawking's office at Cambridge would have made an ideal choice—with its silent aura of concentration broken only by the sound of the clicking device worked by its slumped central figure in his wheelchair. Around him, computer screens, a mirror from which his intent face stares back at you, and large Marilyn Monroe posters gazing down from the walls.

This mind lost to the world is at home in far reaches of the universe. It has produced some of the most exciting cosmological thinking of all time. Our entire image of the cosmos has been drastically transformed during the Hawking era. The picture he and his colleagues have produced is as imaginative and beautiful as a great work of art. It is also as impossible as a dream, and complex far beyond everyday comprehension. Hawking has produced sensational new ideas on black holes, the "theory of everything," and the origin of the universe.

Yet all this has been questioned by some. Cosmology is the study of the universe—but is it really science? For all its fiendish mathematics, much of it cannot be proved. Is cosmology in any way meaningful or useful? Or is it like a fairy tale, as relevant to our lives as the antics of the ancient Greek gods? Hawking's achievement can be seen either as seminal to our understanding of life itself, or as a vast intellectual undertaking filled with sound and fury but signifying nothing. Read on, and judge for yourselves.

Hawking and Black Holes

Life and Works:
A Brief History
of Hawking

STEPHEN HAWKING was born during the darkest days of the Second World War. His parents had a house at Highgate in north London. The night was rent by the howl of air raid sirens, flickering searchlight beams, and the flash and thud of German bombs.

To ensure a safe birth for their first child, Frank and Isobel Hawking decided to move temporarily to Oxford just before he was born. The Germans had agreed not to bomb

Oxford and Cambridge, with their irreplaceable architecture; in return the Allies had agreed not to bomb the historic German university cities of Heidelberg and Göttingen. As Isobel Hawking remarked: "It is a pity that this sort of civilized arrangement couldn't have been extended to more areas." She was safely delivered of a son in Oxford on January 8, 1942. Coincidentally, this happened to be the anniversary of Galileo's death, which had occurred exactly three hundred years before, in 1642. By another coincidence, Newton had been born around the same time that same year. The astrological omens for an astronomer were excellent—if one discounts the fact that these two fields are mutually exclusive.

Both Frank and Isobel Hawking had studied at Oxford. Frank had become a doctor involved in medical research, often abroad. Isobel's career, on the other hand, simply petered out through lack of opportunity—beginning as a bored inspector of taxes, and descending to various unfulfilling secretarial jobs. She was just too early. A few years later Maggie Thatcher took over the Oxford University Conservative Association. During the war, women entered the ministries, riding high in the civil service. Others escaped from domestic serfdom to work as "land girls" on farms, or tasted independence in the factories working at men's jobs.

It was as a secretary that Isobel encountered Frank Hawking, who had just returned from a spell of medical research in Africa. They were soon married, and Isobel was eventually to have four children. She remained very much her own person, and her aim in life was to have a formative influence on her children.

Isobel's life remained essentially unfulfilled, however. She found one outlet in idealism. Initially a believer in communism, she soon softened her stance, but remained a committed Socialist. Later she would take part in the early Campaign for Nuclear Disarmament marches from Aldermaston to London, when attempting to save the human race from nuclear self-destruction was regarded as a wildly antisocial activity.

In 1950 the Hawkings moved thirty miles north of London to St. Albans, a pleasant small cathedral town (or suffocating provincial backwater). Here Frank was head of the division of parasitology at the National Institute for Medical Research. The Hawkings continued to live an orthodox intellectual life, which immediately marked them as dangerously eccentric. Their house was cluttered with books; the furniture was intended for comfortable use, rather than as a status symbol; the curtains weren't washed and sometimes weren't even pulled at night. Those who made it their business to do so noticed that the family listened to the Third Programme on the wireless (advanced drama and classical music, broadcast to those in exile among the philistines). In his spare time Frank even wrote a few novels (never published, and derided as piffle by his wife). The role models for young Stephen were Bertrand Russell and Gandhi, rather than sportsmen or film stars.

In the summer the family would bundle into their car (a former London taxi) and drive down to their trailer for vacations. The trailer was parked in a field at Osmington in Dorset, near Ringstead Bay. (Needless to say, this was no ordinary trailer: The Hawkings owned an old Gypsy trailer, painted in gaudy "Romany" colors.) The Hawkings were not well off, but they were not poor. Likewise, they appear to have been neither happier nor unhappier than most other middle class families during this drab era of social repression.

Out of this average household, a typically average schoolboy emerged. At the age of ten, Stephen was sent to the best local school: the undistinguished St. Albans School, whose tuition was just over fifty pounds per term, or approximately one hundred fifty dollars. Stephen was a puny, gauche, and physically uncoordinated pupil: a recognizable type who fit in among the usual bellowers, breathless mediocrities, whimperers, and complicated oddities that fill any such school playground.

By now Stephen had become interested in "stinks," and even had his own science laboratory back at home. This soon became a typical schoolboy litter of encrusted test tubes, disjointed remnants of long-extinct experiments, and simple guides to manufacturing gunpowder, cyanide, and mustard gas.

It gradually became evident that Stephen was fairly bright, but was not being stretched by the much-vaunted academic standards of his semi-posh school. He didn't work very hard, and still stood well up the class—but was never first. His brain was sharp, yet he spoke too quickly to be readily intelligible. At home in his den with his few school pals, he took to inventing complicated board games. Playing these seldom took less than five hours, and could sometimes last an entire week during the holidays. Not surprisingly, he soon found himself playing mainly against himself. Both friends and family were struck by his ability to become utterly absorbed in some abstruse problem, often for hours on end, until he had finally solved it. In the opinion of his mother: "The game was almost a substitute for living as far as I could make out."

Stephen appeared to enjoy living in a theoretical ordered world, and attempting to challenge its structure to its ultimate limits. He may not have appeared unhappy, but he was certainly not ordinary. His mental focus was unusually abstract, and appeared to be driven by a stronger than natural inclination.

The prizewinner in Stephen's class, his pal Michael, regarded him in friendly dismissive fashion as "a bright little boffin." One day they began talking together in Stephen's lab about "life and philosophy."

Michael reckoned he was pretty hot on philosophy, but as the conversation continued he became aware that Stephen was egging him on, covertly encouraging him to make a fool of himself. It was an unnerving moment for Michael, who suddenly had the feeling that he was being looked down on from a great height by an amused but detached observer. "It was at this point that I realized for the first time that he was in some way different and not just bright, not just clever, not just original, but exceptional." He was also aware of an "overarching arrogance, if you like, some overarching sense of what the world was about." The bright little boffin had evidently spent quite a bit of time thinking about things: attempting to puzzle out what the world was all about.

This was the task philosophy had originally set itself: cosmology. The ancient Greek word for the universe was *kosmos,* a word that also meant "order." The word cosmetic comes from the same base. For the ancient Greeks, the order of the world was a thing of beauty.

Nowadays cosmology has shed its fuzzy philosophical edges, and is limited to the study of the structure of the universe. But the discovery of order in this all-but-infinite vastness can still evoke a sense of beauty and philosophic wonder. This can happen especially in the mind of a pensive, exceptionally perceptive teenager who is drawn to abstraction and capable of extreme concentration in his determination to think to the bottom of things.

Hawking's hidden talents needed a jolt before they would emerge into the light of day. This happened when he was sixteen and studying for his A levels. In 1958 Stephen's father was given a research post in India. The family decided to make an adventure of it and drive there in the car (a daring expedition indeed at the time). But there was one big disappointment: The expedition wouldn't include the entire family. Stephen would have to remain behind and take his A levels, and would be boarded with that nice nearby family, the Humphreys.

The attitude of Mrs. Hawking was all very English. "He had a very good time with the Humphreys, and we had a marvelous time in India." And so it seemed. Though there was a telltale increase in Stephen's clumsiness. On one slapstick occasion the Humphreys lost a trolley full of their best crockery. Recalled Mrs. Humphrey: "I suppose everybody laughed, but after a pause Stephen laughed the loudest."

Whatever else the effect of being abandoned by his family, this was sufficient to spur Hawking's intellect into life. His father had wanted him to study biology, with the aim of following him into the medical profession. Stephen was more interested in mathematics, which he was best at—but his father looked upon this as a dead end, leading only to teaching. Eventually they compromised: Stephen studied math, physics, and chemistry. He applied himself to his A level studies, and also had a trial run at the Oxford entrance exam, with the aim of taking the test for real the following year. Unexpectedly Stephen did so well in the Oxford exam that he was awarded a scholarship on the spot.

At the age of seventeen, Stephen Hawking arrived at University College, Oxford, to study natural science with the emphasis on physics. This lack of mathematics was not indicative of further compromise. On the contrary, Hawking had now come to regard math as only the key to understanding the universe at large. The cosmos itself remained his deepest preoccupation.

Many of the other freshmen were around one and a half years older than the seventeen-year-old Stephen, and the rest were as much as three years older, having completed two years national service. The bespectacled puny-framed Stephen felt young, gauche, and left out of things. He spent most of his first year in his rooms—not working, just being bored and wondering how to get himself accepted by the others. He was too young even to go to the pubs. In the evenings he took to quietly drinking his way through a crate of bottled beer in his rooms while devouring sci-fi. This introduced him to various imaginative, zany, and often woozy views of the universe, but hardly stimulated his academic interest. He was lucky if he put in as much as an hour's work a day.

Hawking's interest was focused on the larger world around him, and this he did study intently, even to the extent of making night observations. He couldn't help noting its singular properties, the intriguing way it behaved and its exciting possibilities. By the beginning of his second year Hawking was ready to enter this world. He'd grown his hair daringly long (for the 1950s), developed a polished wit, and dandified his appearance somewhat. The ugly duckling blossomed, swanning from party to party, taking to the social swim with the self-assured ease of a well-rehearsed mirror-performer. He even joined the hearty heavies of the rowing club, becoming cox of his college eight.

When Hawking set himself to do something, he did it with utter determination. Once again, he seemed to have applied that "overarching arrogance . . . some overarching sense of what the world was about"— which had so shocked his school friend Michael, indicating something exceptional in his character. But this frightening quality was not so much "overarching arrogance," more the self-confidence induced by a focused will.

Yet the focus of this will remained narrow. Hawking wasn't stretched by his studies, and still put in only an hour's work a day. Despite this, his physics tutor Dr. Robert Berman remembered: "He was obviously the brightest student I ever had." Adding: "I'm not conceited enough to think that I ever taught him anything." Such fulsome comments show the hallmark of being made in hindsight. Yet there's little doubt that Hawking was regarded as exceptional, if only because he appeared to defy the conservation of energy principle (the amount you get out of something cannot exceed the amount of work you put into it).

Hawking was full of himself—both socially and intellectually. He saw no point in disguising his exceptional mental ability: Such arrogance only added to one's kudos. Despite his work record, he decided he wanted to continue his studies, doing postgraduate research in cosmology. So he applied to Cambridge to study with Hoyle, the greatest cosmologist of the day, and was accepted on condition he got a first (i.e., a first-class honors degree). No problem.

Not until the very last moment did Hawking's self-confidence fail him. He spent a sleepless night on the eve of his finals, and consequently botched a number of answers. His eventual marks were on the borderline between a first and a second. As was usual in such cases, he was summoned for an interview to decide his fate. By this time his characteristic self-esteem had returned. When asked about his plans, he replied: "If I get a first I shall go to Cambridge. If I receive a second I will remain at Oxford. So I expect that you will give me a first." According to Dr. Berman: "They were intelligent enough to realize they were talking to someone far cleverer than most of themselves." Hawking got his first, and in the autumn of 1962, at the age of twenty, arrived at Trinity Hall, Cambridge.

His arrival at Oxford had been bad enough; his arrival at Cambridge was far worse. For a start, he found that Hoyle had decided not to take him on after all. Hoyle's assistant had been delegated to act as his supervisor. Hawking's pride took a knock: This was a slight he would not forget. In postgraduate Cambridge, Hawking was no longer a star undergraduate. Cambridge had *real* scientific stars, and was used to major scientific events taking place. Crick and Watson had discovered the structure of DNA at the Cavendish

Laboratory in Cambridge, and they were awarded the Nobel Prize within weeks of Hawking's arrival. At the same time, Kendrew and Perutz, also of the Cavendish (and still in residence), picked up the chemistry Nobel. Even in the small world of the Department of Applied Mathematics and Theoretical Physics (DAMTP), Hawking soon began to find things hard going. One hour of study a day had left little time for background work, and his lack of a thorough mathematical grounding soon became evident.

But this was only the visible tip of the iceberg. During his last year at Oxford Hawking had fallen down a flight of stairs and hit his head. As a result, he had suffered a brief memory loss. His friends suspected a drunken escapade. But this wasn't the only time he had fallen down stairs. And on occasion he had also found difficulty in tying his shoelaces. Hawking learned to be careful on stairs, but the latter symptoms persisted.

When he arrived home from Cambridge

at the end of his first term, his father decided he should go into hospital for a checkup. The result was beyond anyone's worst nightmares. Hawking was diagnosed as suffering from amyotrophic lateral sclerosis (ALS), better known as Lou Gehrig's disease.

ALS is a progressive degenerative disease of the nerve cells in the spinal cord and the brain. These cells control muscular activity, and as the disease progresses, the muscles atrophy, resulting in immobility and finally even speechlessness. The body is reduced to a vegetative state, but the mind within it remains utterly clear and functional throughout. Meanwhile all communication becomes impossible. Death usually occurs within a few years. In the final stages the patient is treated with morphine to counteract the effects of chronic depression and terror.

Hawking's reaction was typical of his upbringing and his character. "The realization that I had an incurable disease that was likely to kill me in a few years was a bit of a shock.

How could something like that happen to me?'' His mother's reaction was less understated. She demanded to see the top specialist at the London Clinic. But he grandly informed her: ''There's nothing I can do really. More or less, that's it.''

Despite Hawking's brave words, he was in fact deeply affected. A girl who had met him at a New Year's party, just before he went into hospital, had been somewhat overawed by this scruffy opinionated egghead. When she saw him afterward, ''he was really in quite a pathetic state. I think he'd lost the will to live.'' Hawking returned to Cambridge, and sank into a state of morbid depression. For several months he barely left his digs. All that emanated from his room were the roar of Wagner records and empty vodka bottles.

But gradually the clouds of tragic self-pity began to clear. The girl who had met him at the New Year's party came to see him in Cambridge. She was only eighteen, and named Jane Wilde. She was studying for her A levels at St. Albans High School, and planning to go to London University later in the year.

Jane was shy. When Hawking had first told her that he was studying cosmology, she'd needed to look it up afterward in the dictionary. (Geniuses didn't *explain* such things.) Jane believed in God, and was naturally optimistic. Everything had a purpose; and no matter how bad things might appear, something good could always come out of them. Hawking had long ago dispensed with any belief in God, but Jane's attitude struck a chord. He was strong-willed, he had always been strong-willed: This had been his secret of success. Why should he change now?

"Before my condition had been diagnosed I had been very bored with life," he remembered. "There had not seemed to be anything worth doing." But now things were different. "I dreamt that I was going to be executed," he recalled. "I suddenly realized there were a lot of worthwhile things I could do, if I were reprieved." He was on the mend—mentally, at any rate. Physically, the outlook was not so good.

ALS does not progress in a regular fashion. Each heightening of the symptoms is usually followed by a leveling out, a stabilization that can sometimes last for a surprising time. The doctors had informed Hawking that his disease had entered one of these "plateau" periods, but their prognosis turned out to be faulty. The disease continued to progress, and after a few months Hawking was forced to use a cane to get around. The doctors now gave him less than two years to live. There didn't seem to be much point in starting a Ph.D. thesis, if he would be dead before he had a chance to complete it.

Hawking continued to see Jane, but refused to allow any hint of sentimentality to enter their friendship. He abhorred pity, and was determined to remain as independent as possible for as long as possible. He felt like a normal human being, and he wished to be treated as such. He regarded Jane as "a very nice girl," and she quietly admired his courage. It was this mutual admiration, rather than sentimentality, that made them understand that the impossible might be possible. In Jane's words, they both came to the realization "that together we could make something worthwhile of our lives."

Eventually they became engaged. For Hawking, this "made all the difference." He now had something to live for. But if he was going to get married, he needed a job. And if he was to get a job, he needed a Ph.D.

Hawking's self-confidence returned, and he began thinking about a suitable problem for his Ph.D. thesis. He considered himself lucky. Cosmology required no apparatus other than telescopes, and involved no experiments requiring physical or manipulative expertise. The only thing he *absolutely* needed was his brain, one of the few parts of his body that would remain unaffected by his disease.

In 1965, at the age of twenty-three, Hawking started his Ph.D., and in July he married Jane. That autumn Jane left for London to complete her final year at university, returning to Cambridge at the weekends. Hawking moved into a small terraced house just one hundred yards from the Department of Applied Mathematics and Theoretical Physics, and spent some of the wedding money on a three-wheeled car so that he could drive to the observatory outside town.

Hawking's formidable will was motivated, his mind utterly focused, with the minimum of distraction. And they needed to be. For the problems he now set about addressing were among the most complex and ambitious in all cosmology.

For many years cosmology had been regarded as something of a pseudo-science, and as such had naturally attracted a good number of pseudo-scientists. Big ideas about the universe, backed by improbably large numbers, had succeeded in catching the public eye (and baffling the public brain). Such ideas were the dinosaurs of modern science: enormous, simplistic, and due for extinction. Few penetrating questions were asked. Real scientists preferred real science, which could be proved or disproved by experiment. The bamboozled public was only expected to gasp in awe at the latest news about the universe. Objections were not required.

By the early 1960s all this had begun to change. The great discoveries of the early twentieth century—relativity and quantum theory—had transformed our view of both the subatomic world and the universe. Relativity meant that space was curved and the universe had a boundary. But only now were relativity and quantum theory being rigorously applied to the nitty-gritty of the universe, on both the subatomic and the galactic scale. What effect did these ideas have on the vast, continuous experiment which *made up* the universe? The answers were—and continue to be—wilder than the wildest imaginings of science fiction. Who could have conceived of black holes, invisible gaps in the universe where space and time simply disappeared?

Hawking had noted that relativity did not accord with physics at the quantum mechanical level, and was therefore insufficient to explain black holes. His investigation into what this implied was to produce a sensational result.

Astonishingly, the existence of black holes (though not called such) had been predicted as long ago as 1783. This had been done by an English village parson, John Michell, who also happened to be one of the finest astronomical thinkers of his day. (Besides black holes, he also suggested the nature of double stars, and made some remarkably farsighted predictions of stellar distances.)

Michell suggested that if a star was sufficiently large and dense, no light would be able to emanate from its surface. His observations of the heavens led him to theorize that the universe contained a considerable number of such stars, whose presence could be detected by their gravitational effect on nearby visible stars or planets.

This idea was revived in the early years of the twentieth century by the German astronomer Karl Schwarzschild. During a spell on the Russian front in 1916, he began working out the implications of Einstein's recently published general theory of relativity. This had theorized that light rays could be bent by gravitational attraction. (Life on the Russian front was almost as dangerous and uncomfortable as life in the trenches on the Western front, but there must have been something intellectually stimulating in the air: At the very same time, further down the line, the Austrian Ludwig Wittgenstein was thinking out the ideas that were to transform twentieth-century philosophy.)

Schwarzschild demonstrated that certain things will happen when a star collapses under the force of its own gravity. According to Einstein's theory about the effect of gravity on light, after a certain point the gravitational force effect will increase to such an extent that nothing, not even light, will be able to escape from its gravitational field. This point will be reached when the star collapses to a certain radius, dependent upon its mass. This radius is the point where a collapsing star becomes a black hole. In the case of the sun, whose present radius is 700,000 kms, this would become a black hole if its radius contracted to 3kms (8 kms = 5 miles). Schwarzschild had proved by relativity what Michell had only suspected.

Curiously, Einstein refused to accept Schwarzschild's findings—even though they were based upon his theory. Nonetheless, the critical radius at which a star becomes a black hole is now known as the Schwarzschild radius.

A year later Einstein found his cosmological ideas contradicted once more, this time

by the Russian astronomer Aleksandr Friedmann, working in St. Petersburg, then called
Petrograd. While the Russian Revolution was
taking place outside his window, Friedmann
worked out that Einstein's picture of a static
universe was incorrect. In the course of his
calculations Einstein had assumed a "cosmological constant," which he had named
lamda. This in fact begged the question of
whether the universe was static. Friedmann
showed that there was no justification in making such an assumption.

Friedmann took the daring step of assum-
ing that the universe was filled with a uni-
formly thin cloud of matter. (Modern findings have confirmed that this daring
assumption holds good for many macrocos-
mic calculations, despite the obvious discrepancies.) Working from this model, and a
suitably modified version of Einstein's calculations, Friedmann was able to show that the
universe must in fact be *expanding*. Once
again, Einstein chose to differ.

Friedmann's theoretical assumptions were confirmed by practical observation in 1928 by the American astronomer Edwin Hubble (after whom the space telescope is named). Unaware of the theories of either Einstein or Friedmann, Hubble began studying the red shifts of over a dozen different galaxies, using the one hundred-inch telescope at Mount Wilson. (The red shift is a displacement of lines in the spectrum indicating speed relative to the observer.) Hubble discovered that the speed at which these galaxies were receding became greater the farther they were from the earth. This was the first practical evidence of an expanding universe.

The next major theoretical advance was made five years later, and also came from Russia. By now Stalin's purges were in full swing. It may have been possible for a dedicated scientist to ignore the Russian Revolution going on outside his window, but Stalin's Terror was a different matter. Men in bulky leather overcoats knocked at the door and demanded to be let in—even if you were extremely busy with cosmological calculations. After the generals and the party bosses, top scientists were now in demand for starring roles in the show trials.

The theoretical physicist Lev Landau knew he was in deep trouble: Not only had he recently returned from working abroad, he was also Jewish. Landau decided his only hope was to achieve such worldwide eminence that his appearance on the witness stand (and consequent disappearance) would prove an embarrassment to the Soviet utopia. Rapidly he dashed off a paper containing some sensational cosmological ideas that he had been pondering for some time. This he sent posthaste to his pal, the great physicist Niels Bohr, in Copenhagen. In his accompanying letter, Landau asked Bohr a favor. If he found the paper any good, could he use his influence to get it published in *Nature*, the top international scientific journal.

A short time later Bohr received a telegram from the official party newspaper *Izvestia,* demanding to know if Landau's paper was any good. Bohr had not had time to study the paper, but quickly got the picture. He sent a message of fulsome praise back to Moscow, and made sure that Landau's paper was published in *Nature.* (Despite this, Landau was arrested in 1938—but was soon released, as this was discovered to have been a "mistake.")

Landau had been speculating for some years about how stars produced sufficient energy to account for their great heat. In his *Nature* paper he theorized that the center of a star consisted of another, super-dense star made largely of uncharged subnuclear particles known as neutrons. (A star such as the sun would contain a neutron star around one-tenth of its mass, but this would be compressed into a radius of just 1km.) The inordinate heat emanating from a star was generated by the inner neutron star's absorption of gas.

Landau's paper had been written in some haste, and was published before he had found time to think his ideas through properly. This paper was read by the ace American quantum physicist Robert Oppenheimer and his brilliant assistant, Hartland Snyder, who had earlier worked as a truck driver in Utah.

Oppenheimer and Snyder found many shortcomings in Landau's paper, but built on his original idea. According to Oppenheimer and Snyder, when a large star exhausted its nuclear fuel and burned out, it then imploded under its own gravitational attraction. At a certain point it contracted to a critical radius, where even light rays were unable to escape from its surface. At this point the star became isolated from the rest of the universe, and a "one-way event horizon" developed. Particles and radiation would be able to enter, but nothing would be able to escape. A space-time singularity would be formed, where the dimensions of space, and its linked dimension, time, simply disappeared. There could be no way of telling what went on inside this horizon, and Oppenheimer refused even to speculate on it.

Oppenheimer and Snyder made public their findings in the *Physical Review* on September 1, 1939. This was the day Hitler invaded Poland, precipitating the Second World War. In the same issue of the *Physical Review* Niels Bohr and the American physicist John Wheeler published an article on how to obtain nuclear fission (i.e., the mechanism necessary to produce an atomic bomb). By coincidence Oppenheimer was to head the Manhattan Project, which developed the first atomic bomb. On the very day that the Second World War began, the method by which it would be brought to an end was published—alongside a paper by the man who would make this possible. But for the time being Oppenheimer's paper was largely ignored: The world now had something more important to worry about than the universe.

Wheeler was eventually to work on the hy-

drogen bomb, but when he had finished
working out how to destroy planet earth, he
turned his attention to the universe. Fortunately cosmology was concerned with holism
rather than holocausts, though Wheeler still
managed to introduce certain unfinished
business from his previous field. Wheeler was
a right-wing extremist, an orthodox American position in the 1950s era of McCarthyite
anticommunist witch-hunts. Oppenheimer,
on the other hand, had once slept with a
Communist—which meant that despite winning the war by producing the A-bomb, he
was of course a communist spy. Wheeler
didn't approve of Oppenheimer's cosmological ideas either, but was eventually forced to
concede that there might be something in his
idea of a space-time singularity existing
within a one-way event horizon. Indeed,
Wheeler was to go even further and christen
this "gravitationally completely collapsed ob-
ject": He was to call it a "black hole." Perhaps inevitably, Wheeler couldn't agree with

45

all that Oppenheimer had said. Wheeler maintained that it *was* possible to describe what would happen in a black hole. There would take place a merging of relativity and quantum physics.

But in the early 1960s many still doubted the very existence of black holes. Indeed, Wheeler's worst political suspicions must have been confirmed when a group of Soviet scientists announced they had proved that space-time singularities (black holes) simply couldn't exist. According to the Soviets, such space-time singularities were simply a mistaken theoretical conjecture that only arose if one assumed that large collapsing stars imploded in a symmetrical fashion. Only in this way would the concentrating gravitational field focus on a single point, causing a space-time singularity. Without this unlikely symmetry, there would be no singularity. Abracadabra: no black holes.

As we can see, cosmology in the early 1960s, when Hawking entered the scene, was in a highly fluid state. Indeed, the prevailing orthodoxy at Cambridge was still in favor of the steady-state theory proposed by Hoyle. According to this, the universe had not begun, and would not end, it had always existed—its overall mean density remaining forever constant (i.e., in a steady state). It was Hoyle who in the 1950s had dismissively dubbed the opposing notion the "big bang" theory, ridiculing its idea of creation as "a party girl jumping out of a cake."

Yet Hoyle's steady state theory required similar conjuring tricks. How could he explain away the expansion of the universe, which had actually been observed by Hubble? To overcome this minor point, Hoyle proposed that stars and galaxies were in fact continuously being created out of space. But how? According to Hoyle, this was simply one of the properties of space. (And to make up for this, stars and galaxies were also continuously disappearing into the wide black yonder.)

Hoyle was a tireless and sometimes overhasty self-publicist in the cause of his steady state theory. On one famous occasion he delivered a speech at the Royal Society in London before he had done the calculations to support his assertions. Unknown to Hoyle, Hawking had been shown the preliminary figures by Hoyle's assistant, and had spotted a number of anomalies. Hawking decided to attend Hoyle's speech at the Royal Society, which was received with enthusiastic applause. Hoyle then asked if there were any questions. A frail bespectacled young graduate student struggled to his feet with the aid of a cane. The hundred-strong audience, which included many distinguished scientists, turned to examine this newcomer who had the temerity to question the famous man.

"The quantity you are talking about diverges," said Hawking.

An excited murmur arose from the audience: If this was the case, Hoyle's speech was nonsense.

"Of course it doesn't diverge," replied Hoyle dismissively.

"It does," insisted Hawking.

"How do you know?"

"Because I worked it out," stated Hawking evenly.

A few titters arose from the audience. Hoyle was incandescent with rage. Who was this bumptious young upstart?

Hawking had announced his arrival on the cosmological scene with a vengeance.

But the problem of what happened inside a black hole still remained. Those who took the nonsymmetrical view of collapsing stars, similar to that maintained by the Soviets, had begun to develop a new picture. According to this, the star would implode so unevenly, and so powerfully, that it would simply "fly past itself" and expand again.

This problem was tackled by a young British mathematician named Roger Penrose. He applied his newly developed mathematical methods in topology to the problem of collapsing stars, and came up with some intriguing results. According to his singularity theorem, the collapsing star would behave just as Wheeler had predicted. It would form a singularity where time ceased and the laws of physics no longer applied. And even if it imploded unevenly, the matter would not then fly past itself to expand again. A large collapsing star would implode to its event horizon, where it would become a black hole. (For a star ten times the size of the sun, this would happen when its radius shrank to 30km.) But Penrose established that beyond this point the collapsing star would *continue* to shrink. It would do this according to the picture already established by the general theory of relativity. As the gravitational field intensified, all light, matter, and space-time would continue being drawn into it with ever-increasing intensity. Indeed, it would continue to shrink with such growing intensity that it would eventually have zero volume and infinite density. In other words, it would defy the laws of gravity to the extent of having mass but no dimension. Likewise, space-time and light wouldn't just be drawn into a hole; they would be wound up infinitely tightly to the point where they disappeared.

All this would happen *within* the event horizon, and would thus remain unobservable. But the event horizon would not shrink or implode in any way: It would remain the same—at the point when the imploding star became a black hole. (For example, the event horizon for the star ten times the size of the sun would be maintained at a radius of 30kms while within this the star itself shrank to infinite smallness and density.)

Hawking began studying Penrose's ideas in some detail, and as he did so an idea of astonishing originality began to form in his mind. Like many such great ideas, this was essentially simple (though the mathematics to back it up proved anything but). Hawking asked himself what would happen if a black hole could somehow reverse itself. He then applied this idea to the entire universe. What if the expanding universe was no more than a huge collapsing star *in reverse*. Time vanishes *into* a black hole: If this process were reversed, it would involve the creation of time. Likewise space. Matter would originate from an infinitely dense but dimensionless point. And this point would be the big bang—the very act of creation, no less.

The theory of relativity applied *both ways*. As the gravitational field intensified, space-time, matter, radiation, were concentrated. As the gravitational field expanded and weakened, so space-time unfurled, radiation and matter spread. Hawking succeeded in demonstrating that there must have been a singularity in the distant past that had originated time. And if the universe stopped expanding and began to contract, it would eventually implode and *end* in a singularity—the so-called "big crunch." There was no question of what happened before the universe began, or what happened after it ended—for under such circumstances there was *no such thing as time.* Space too would be nonexistent, along with matter.

Hawking had explained how the universe originated. He had shown how the big bang actually worked, how it had come about from an all-embracing reverse black hole. (Though the Soviets gamely continued to maintain that there was no such thing as a black hole, and Hoyle doggedly continued to defend his steady-state theory.) Word of Hawking's astonishing theory soon began to spread, gaining wide acceptance in all but the Soviet and flat-earth universe. Hawking had established himself as a rising star on the cosmological scene.

But cosmology remained a small world, and Hawking's fame was confined to matters relating to the universe. In the larger world of Cambridge academia, he was merely a peripheral genius figure (and one of many such). Yet the legend was beginning to grow. The graduate students in the DAMTP building had become used to encountering the frail, bespectacled figure with his cane, who brusquely refused all offers of assistance. Often he would stand for minutes on end, gasping against a wall as he fought to climb the stairs. It was now four years since he had been given two years to live, and he was increasingly forced to resort to crutches. These he detested: They not only marked him out as disabled, but also seemed to exhaust him even more.

Yet Hawking remained very much himself, and his body was still in a far from useless condition. In 1967 his son Robert was born; and despite encumbrances such as crutches, Hawking devoted long arduous hours to his work. He was filled with enthusiasm for what he was doing. Ironically he now felt happier than he had been before his illness, or so he maintained.

But none of this would have been possible

without the constant and selfless support of his wife Jane. Life was not easy with a "more or less human genius" who was possessed of the usual emotional extremities associated with this type. Tantrums were not infrequent, and Hawking remained more than capable of expressing the full force of his personality. Though he might be a genius and disabled, he insisted upon being treated as a human being. And despite the difficulties, this was still possible. His marriage was close, and not entirely separated from his work. Jane typed up his papers from his scrawled notes, or took dictation from his increasingly frail voice. His speech was beginning to degenerate into a slurred moan.

Hawking now carried out more and more of his mathematical work mentally, training himself to achieve an exceptional cerebral dexterity that accommodated this. Increasingly, he took to communicating this intellectual work only when it was in developed form. The powers of memory, concentration, and mental organizational skills required were formidable. To say nothing of the sheer willpower involved. And this was just the back-up work. On top of this was required the creative power and insight to produce original thought of the highest order. And this he continued to do.

As Hawking's renown began to spread, he attracted a team of highly talented fellow researchers to DAMTP, who collaborated with him in his continuing investigation into black holes. In 1971 Hawking came up with the idea that after the big bang a number of "mini black holes" had formed. These were so concentrated that they contained a billion tons of matter, yet were no larger than a photon, the elementary particle in which light is emitted. Hawking demonstrated that these mini black holes were unique—owing to their enormous mass and gravity, they would obey the laws of relativity, yet their minuscule dimensions required that they also followed the laws of quantum mechanics. This suggested that "in the beginning" these two often con-flicting explanations might have been one. It gave a hint that perhaps, in the not-too-dis-tant future, it might be possible to develop an overall theory that accounted for both quan-tum mechanics and relativity. Yet for the time being such sensational possibilities remained not even remotely realizable.

Indeed, precisely the opposite was the case. A singularity produced by gravitational collapse meant the breakdown of all the known laws of physics. Shock, horror, deprav-ity! But because this event took place within a black hole, it was unobservable; we were pro-tected from viewing such ultimate obscenity by a form of "cosmic censorship." Yet if the laws of physics broke down, this meant that it was impossible to predict what might happen in the future. In which case, science had a huge hole blown in it.

Philosophically, science was now faced with two sensational and conflicting possibili-ties, both of which could be called "the end of science.'' Mini black holes hinted that there might one day be a theory that explained everything. At the same time, more commonplace black holes indicated that the universe might simply not be amenable to scientific explanation—ultimately, it might not be scientific at all. Science had now reached the ultimate philosophical stage. It was living dangerously—what lay ahead was the possibility that it would either be completed, or exploded. The end of science was at hand!

But science wisely tends to ignore such philosophical quibbles. Regardless of this imminent demise, Hawking and his fellow cosmologists persisted in their research. It might have been impossible to see inside black holes, where the laws of physics no longer applied, but it was always possible to *conjecture* what happened within this forbidden territory. Their origin had been explained—now it was a matter of explaining their continuing existence.

Across the Atlantic, Wheeler had not only

christened black holes, but had come up with just such a conjecture, known as the "no-hair theorem." According to this, a black hole soon achieves a stationary state where only three parameters hold good. Namely: mass, angular motion, and electric charge. When anything enters a black hole, only these three entities are conserved.

By 1974 Hawking and his team had managed to prove the "no-hair theorem." (Liken the "hair" to the protruding coordinates of the dimensions, and other adhering physical fuzz, which is shaved off on entering the black hole—so that only bald, electrically charged, mobile mass makes it inside.) Hawking showed how relativity could account for Wheeler's conjecture. The laws of physics might break down inside a black hole, but it was not complete anarchy in there.

During the academic year 1974 to 1975 Hawking took up an invitation to spend a year at Caltech. This was the most prestigious scientific establishment on the West Coast—

where the twentieth century's finest chemist Linus Pauling had worked, now home to a cluster of Nobel laureates. (Among whom were such scientific illuminati as the bongo-playing physicist Richard Feynman, and Murray Gell-Mann who was liable to name his discoveries after a quote from James Joyce or a Buddhist text.)

Hawking enjoyed California, taking the opportunity to use the powerful telescopes at Mount Wilson, and successfully discouraging anyone from taking him to Disneyland—though he did acquire a large poster of Marilyn Monroe, which became the first of several to adorn his office back at Cambridge.

By now Hawking's ALS had arrived at another "plateau," but in reaching this it had reduced him to a wheelchair. Likewise, his voice was beginning to deteriorate to a barely intelligible moaning noise, such that only close colleagues and friends were able to understand him. In defiance of such crushing disabilities, Hawking became a father for the third time in 1979. As one of his more forth-right friends publicly announced several years later when introducing Hawking at a lecture: "As evidenced by the fact that his youngest son Timothy is less than half the age of the disease, clearly not all of Stephen is paralyzed!" The audience froze in uncertain embarrassment, but the small twisted figure in the wheelchair produced his famous broad grin.

At the age of thirty-two, Hawking had been elected to the Royal Society, one of the youngest ever fellows. Other prizes and honors began to follow. According to his long-suffering wife Jane, these awards were "like the sugar frosting on the cake." But life with Hawking was not easy for her: "I don't think I am ever going to reconcile in my mind the swings of the pendulum that we have experi-enced in this house—really from the depth of a black hole to all the glittering prizes."

It was around this time that Hawking had his famous "eureka moment," which set him on the path to his major discovery. One evening as he was going to bed he began thinking about the surface of black holes. Hawking's obstinate insistence on doing everything for himself meant that going to bed was for him a lengthy and laborious process—so he had quite a bit of time to think as he went about it.

Hawking began pondering what happens to the light rays at the event horizon of a black hole. He knew that the rays of light that form the event horizon, the surface of the black hole, can never approach one another—because they are held in suspension, neither able to escape nor be sucked into the black hole. In a sudden flash he realized what this meant. *The surface area of a black hole can never decrease.* In other words, even if two black holes combined, they wouldn't swallow each other up. On the contrary, their total surface area could only stay the same or increase, it could never decrease. This may seem a rather abstruse point—one that is neither particularly exciting nor significant. Yet its implications were to change our entire notion of what a black hole is. Hawking sensed this, and the excitement of it rendered the arduous task of getting himself into bed redundant. He spent a sleepless night.

Hawking had realized that the behavior of the surface of black holes bore an uncanny resemblance to the second law of thermodynamics. This states that the entropy (or disorder) within an isolated system will always stay the same or increase; and if two such systems are joined, their combined entropy is greater than the sum of the previous entropies. Basically this means that if things are left to themselves, disorder will remain the same or increase. It can never decrease. (Hawking himself gave the example of a house. If you stop repairs, the disorder increases. To create order, or repair disorder, a further input of energy is required.)

This law accounts for why certain processes are irreversible. If you drop a glass, it cannot put itself together again—this would decrease its entropy, if we regard the glass as a separate system. Entropy determines the direction in which an irreversible process must go. In a way, it indicates the direction in which time must proceed.

So why did the behavior of black holes echo the second law of thermodynamics? Could it mean that this law somehow applied to black holes—which had previously been regarded as entities where such laws no longer applied?

Up till now, calculations concerning black holes had been based upon relativity, which accounted for the behavior of large bodies. Effects taking place at subatomic level, which conformed to quantum theory, had been discounted. Minuscule subatomic effects would be inconsequential when dealing with magnitudes as huge as collapsing stars and black holes. Hawking was to show how wrong this assumption had been. Quantum mechanics provided a vital clue to the true nature of black holes.

First it is necessary to understand a little about quantum mechanics. One of the most fundamental and intriguing notions of quantum physics was put forward in 1927 by the German physicist Werner Heisenberg, when he was just twenty-six years old and already a major exponent of quantum theory. Heisenberg's great discovery was the uncertainty principle, which states that it is impossible to determine simultaneously both the precise position and the precise momentum of a particle.

Heisenberg maintained that this cannot be done, even theoretically, because the very notions of precise position and precise velocity, taken together, have no meaning in nature. (This is in fact true of *all* things in nature, from subatomic particles to giant tortoises to galaxies—but only at atomic levels and below do the discrepancies involved become significant.)

A simple illustration of this is given if we try to determine the precise position of an electron. This particle is so small that it can be detected only by something of sufficiently small wavelength, such as gamma rays. But when these rays hit the electron, they affect its momentum in an unpredictable fashion. It is impossible to determine the electron's position without altering its momentum. And the more precisely we try to determine its position (using shorter waves), the more this will affect its momentum. Likewise, the less we interfere with its momentum, the less precisely we can measure its position.

As for particles, so for fields—which may be regarded as consisting of particles. Heisenberg's uncertainty principle yields astonishing results when it is applied to space:

—Space too is a field.

But how? Surely space is by definition empty, a vacuum.

—According to Heisenberg's uncertainty principle, this simply cannot be the case.

Why not?

—We have shown that it is impossible simultaneously to measure the value of a field, and the rate at which it is changing, with absolute accuracy. This is true for fields, just as it is for particles.

So?

—This means that no field can measure precisely zero. This would be an *exact* measurement of both its value and its rate of change. Impossible, according to the uncertainty principle. Yet if we are to have empty space, this field must be precisely zero.

So there's no such thing as empty space?

—Precisely. (Or perhaps almost precisely!)

So what do we have instead?

—According to Heisenberg's principle, even in space there will always be the tiniest uncertainty. But what does this mean?

—This uncertainty can be imagined as a minuscule oscillation from just above zero to just below zero—but never *actually* zero.

And how does this happen?

—We have to account for what happens in the following way. We can't have nothing, so instead we have pairs of virtual particles. These account for the oscillations either side of zero.

But what are these virtual particles, and how do they account for the oscillations?

—The pairs of virtual particles consist of a particle and an antiparticle. One is positive, and one is negative. When they come together they annihilate one another. These pairs of virtual particles are constantly bobbing in and out of reality, forming and annihilating each other. This accounts for the tiny oscillations above and below zero.

So what has all this got to do with black holes?

—Black holes exist in space, which means that this process is going on all around them.

Hawking speculated on what would happen at the precise surface of a black hole, the event horizon. This space too would contain pairs of virtual particles, popping into reality.

But before they could annihilate themselves, they would be affected by the black hole. The black hole would attract the negative particle, while at the same time it would eject the positive particle. This would escape in the form of radiation. The black hole would effectively be emitting thermal radiation (i.e., heat). It would thus have a measurable temperature.

Likewise, the high-entropy particle falling into the black hole would cause the surface of the black hole to increase. (As we have seen, the surface of the black hole conforms to the Schwarzschild radius, which is dependent upon the mass involved.) The increase in the surface of the black hole, be it ever so minute, marks an increase in the entropy of the black hole. But if the black hole has entropy, this also indicates that it must have a temperature.

This temperature would in reality be almost negligible—mere millionths of a degree above absolute zero—but it would undeniably be there. Hawking had shown that black holes were not "black." They emitted radiation—heat, as if they were hot.

The implications of this transformed the entire conception of black holes. They were not, after all, limitless plug holes in the universe, down which matter, space-time, and the laws of physics disappeared. Black holes could now be regarded as objects that exist *within* the universe. They obey the second law of thermodynamics. They have entropy. This meant they even have time. They are no longer invisible—they could be "seen" by the laws of physics.

But this was not all. In combining the gravity of black holes and the behavior of virtual particles, Hawking had in effect combined quantum mechanics and relativity for the first time.

Word soon began to spread that Hawking had come up with some ideas that "changed everything." As a result, in February 1974 Hawking was invited to speak at a conference in Oxford on the subject of black holes. This had been organized by the mathematician John Taylor, who considered himself something of an expert on black holes. After the other speakers had delivered their papers, Hawking was wheeled to the front of the lecture hall. He began speaking in his moaning, barely comprehensible voice. The audience strained to listen, hardly able to believe what they were hearing. If what Hawking was saying was true, this did indeed change everything. Hawking ended by making an even more sensational claim. A black hole had time, it had entropy, and this entropy increased like any other. This meant that eventually a black hole would evaporate into pure radiation. In other words, in the end it would ''explode.''

The audience greeted Hawking's speech in stunned silence. Then Taylor leapt to his feet and declared: ''I'm sorry, Stephen, but this is complete nonsense.'' Barely able to contain his fury, he turned on his heel and stalked out of the room.

A month later Hawking published a paper outlining his findings. It was published in *Nature,* under the title "Black Hole Explosions?" This paper was described by Hawking's former tutor and collaborator David Sciama as "one of the most beautiful in the history of physics." It has been hailed as Hawking's equivalent of Einstein's general relativity paper. Its importance, though fundamental, is not quite of the same magnitude—but it did succeed in producing a similarly antagonistic response from those who refused to understand it. Some months later Taylor published an angry reply in *Nature,* pooh-poohing Hawking's idea of exploding black holes. But by then the battle was virtually over. Taylor's ideas, like Hoyle's steady state theory, were already a thing of the past. The scientific world is not exempt from evolution. Here too the survival of the fittest holds good—even if these don't immediately appear to be among nature's finest specimens.

* * *

By now Hawking's illness had progressed to an alarming degree. He could no longer walk, even aided, and was forced to move around in a motorized wheelchair. He was unable to feed himself, and when his head fell forward onto his chest he wasn't even able to raise it himself. These were deep psychological blows to a proud and willful man who cherished his independence. But there were even more ominous developments. Hawking's speech was continuing to deteriorate—even those who were closest to him would soon have difficulty understanding what he was trying to say. At the same time, he was rapidly losing his ability to write. His mind had now reached the height of its powers—but how would he be able to communicate his thoughts?

Yet what could be expected? It was now *fifteen years* since Hawking had been given just two years to live. His survival at all was miraculous—almost as miraculous as the discoveries he continued to make in cosmology. The link between these two was not fortuitous. Both were indicative of exceptional qualities of mind and will.

In 1979, at the age of thirty-seven, Hawking was appointed Lucasian Professor of Mathematics at Cambridge. This was the most prestigious post of its kind in the land—the chair previously held by Isaac Newton, and later by Babbage, the father of the computer. Hawking felt deeply honored. Several months later, when he realized he hadn't signed the historic register of Lucasian professors, he took great pains to inscribe his signature. As he later remarked: "That was the last time I signed my name."

Despite his difficulties, Hawking insisted on moving in the Cambridge social swim. He and Jane went to restaurants, attended parties, and the new Lucasian professor soon gained a reputation as a popular host. None of this would have been possible without

Jane, who in the words of a close friend was "a remarkable woman. She sees that he does everything that a healthy person would do. They go everywhere and do everything." His biggest regret was not being able to play physically with his growing children. Hawking also began using his new prestige to campaign for the disabled. His combative nature found an enjoyable outlet in aggressive letters to Cambridge City Council, on such matters as the installation of ramps and the lowering of curbs. His success in these campaigns brought him a "man of the year" award from the Royal Association for Disability and Rehabilitation.

Hawking's ALS may have reached a plateau, but many of his theoretical physicist friends now felt that he couldn't last much longer. The end was in sight. Hawking characteristically pulled the carpet from under these friends with his inaugural lecture as Lucasian professor. This was entitled: "Is the End in Sight for Theoretical Physics?" The lecture was attended by a large crowd, and was read out by one of Hawking's students.

Here Hawking launched into a topic that was to become something of a hobbyhorse. Namely, a "theory of everything." This would provide a unified, consistent, and complete description of *everything*. (In this case, all elementary particles and all known physical interactions in the universe—encapsulated in one set of equations.) It would mark the "end" of theoretical physics. Hawking admitted that after this there would "still be lots to do," but it would be "like mountaineering after Everest."

Such an "ultimate explanation" has proved a remarkably resilient delusion. The first ancient Greek philosopher Thales of Miletus, who lived in the sixth century B.C., was convinced that he had found it (water). And through the centuries since, philosophers and scientists have constantly been convinced that they've found it, or been on the brink of finding it. Candidates have included: fire, breath, atoms, the axioms of geometry, monads, gravity, atoms again, logical language, and many, many more. At the time of his Lucasian lecture, Hawking thought there was a good chance that a theory of everything would be discovered by the end of the century (the twentieth, that is). He even suggested a possible candidate in $N = 8$ (supergravity). It had been suspected for some time that a form of gravity might hold the key, since the constant of gravity (G) appeared to determine the structure of the universe, and was perhaps proportional to its age. But in the end this theory proved to be more complex than it was comprehensive.

Hawking has since revised his view in favor of the superstring theory. This claims the fundamental objects that make up the universe are one-dimensional, stringlike objects, rather than minuscule particles. These infinitely thin fettuccine are said to be around 10^{-35} meters long, and may yet unify all known particles and forces in the ultimate bolognese. Even so, Hawking now predicts that the superstring theory will take at least another twenty years to unravel. Then we'll have solved the final problem—it will be possible to know everything.

However, at this point it's worth bearing in mind the words of Wittgenstein, when he thought he had reached "the final solution of the problems" of philosophy. Only then did he realize "how little is achieved when these problems are solved." Unlike science, philosophy has come of age in the twentieth century with the realization that there is no such thing as ultimate truth. Neither in the philosophic sense nor in the scientific sense. Both science and philosophy are just systems by which we live, *and our notion of these systems too evolves* along with our notion of truth. Both these systems are based upon our notion of truth. Both these systems are based upon what is useful to us, and fit in with how we choose to view the world. The ultimate superstring is liable to be no more "the truth'' than fire or atoms. (Or, on the other hand, it will appear *just as true* as they did in their time.)

In defiance of his illness, Hawking still insisted upon traveling. He was now becoming an internationally famous scientific figure, and was determined to take his place on the international scientific stage. Visits to Switzerland, Germany, and the United States took place. Hawking's physical condition now meant that he was increasingly forced to rely on his memory. With typical persistence, he trained this to a phenomenal pitch. In a seminar at Caltech he astonished the assembled students by dictating a forty-term equation from memory. Alas, the quantum wizard Gell-Mann was present, and felt bound to point out that if *his* memory served him correctly, Hawking had missed out a term. It turned out that Gell-Mann was correct. Where there is supergravity and superstring, there is also liable to be supermemory.

In the early 1980s Hawking started dictating some ideas for a popular book about cosmology. He intended earning some money to pay his daughter's school tuition. By 1985 he had finished the first draft, and decided to read through it during his summer vacation. At the time he was staying in a rented apartment in Geneva, being looked after by a nurse and a research assistant, while Jane took her vacation touring Germany. Between editing his manuscript, Hawking spent time at CERN, the nearby European nuclear research establishment. Here the vast particle accelerators (some several kilometers in circumference) were yielding exciting new practical information about subnuclear particles.

One night, when Hawking's nurse looked into his room at 3 A.M. on her routine half-hourly check, she discovered that something was seriously wrong. Hawking was purple in the face and struggling to breathe. A gurgling noise was coming from his throat.

Hawking was rushed to hospital, where he was immediately placed on a ventilator. The doctors discovered that he had a blocked windpipe and was suffering from pneumonia—a common occurrence in the later stages of ALS. For a while it looked as if he might not survive until morning. A succession of panicked phone calls were made to the list of numbers that Jane had left, and eventually she was traced to Bonn, almost four hundred miles away.

By the time Jane arrived that afternoon, Hawking was out of danger, though he was still on a life-support machine. She found herself faced with an agonizing decision. Hawking required a ventilator to breathe. There was practically no chance of him surviving unless he had a tracheotomy—an operation that involved cutting open his throat and inserting a device that would enable him to breathe. This would save his life, but it also meant that he would not be able to speak again. Was she willing to condemn one of the finest scientists of his age to silence for the rest of his life? Jane decided that her husband's life was more important than anything he might have to say, no matter how universe-shaking it might be. Hawking was operated on, and lost all power of speech.

Back in Cambridge, the Hawkings were forced to pick up the pieces. He would now require the most costly round-the-clock nursing, an expense they simply couldn't afford. (The National Health Service had suggested he be confined to a home for incurables.) The only way Hawking could communicate was by blinking his eyes, and laboriously indicating letters on a board held in front of him.

Jane set about writing begging letters to charitable organizations all over the world. Fortunately an American charity soon came up with financial relief. News of Hawking's plight spread through the scientific community. As a result, the Californian computer expert Walt Woltosz sent Hawking a computer program he had just written. Called Equalizer, it allowed him to select any one of three thousand words from a menu on a screen.

This computer setup was fitted to Hawking's motorized wheelchair by his friend David Mason, whose wife Elaine became one of his nurses. The sensor for this machine could be moved by means of a handheld switch, requiring minimal finger movement, all he could now manage. When a sentence had been put together, it was conveyed by voice synthesizer.

All this required practice. But after a while, one of the finest minds of his time was able to manage as many as ten words a minute. (In other words, the preceding sentence would have taken him nearly two minutes, allowing for shortcuts.) "It was a bit slow," commented Hawking, "but then, I think slowly, so it suited me quite well."

The truth behind these words was not so rosy. In fact, he hated the synthesizer. In the kind understatement of his biographers Michael White and John Gribben: "It doesn't sound too much like a robot." And in Jane's words: "There were days when I felt sometimes I could not go on because I didn't know how to cope.''

Meanwhile Hawking continued with the scientific quest for the Holy Grail: the "ultimate explanation." In order to reach this, it would be necessary somehow to combine the four known forces that had so far been discovered in the universe.

These are:

1. Gravity. This controls the larger structure of the universe, including the galaxies, stars, and planets. (Gravity had in fact been put forward as a previous contender for the title by Newton in the seventeenth century—superseding clockwork, as proposed by French and German philosopher-scientists of the preceding generation.)

2. Electromagnetic force. This is the "glue" that holds all atoms together. It also accounts for all chemical reactions.

3. The strong nuclear force. This holds to-

gether neutrons and protons in the nuclei of atoms, and accounts for such reactions as nuclear fission and fusion.

4. The weak nuclear force. This is responsible for the radioactive decay of nuclei, when alpha and beta particles are spontaneously emitted.

These four forces separated to become distinct entities when the universe was less than one nanosecond old. (A nanosecond is one billionth [10^{-9}] of a second.)

As we have seen, ideas along the line of the theory of everything have a long history (almost as long as science itself). However, the theory in its present form only came into its own in the twentieth century—when quantum theory and relativity transformed our view of the universe. At that stage it was thought that only two forces operated in the universe: gravity and electromagnetism.

In the 1920s, parts of Maxwell's electromagnetism were combined with the quantum theory of gravity to become quantum electrodynamics. This was optimistically entitled QED (echoing *quod erat demonstrandum*, meaning "which was to be demonstrated," as appears at the end of a successful geometric proof). QED looked set to explain everything. So much so that in 1928 the professor of physics at Göttingen, the great German theoretician Max Born, was able to announce: "Physics, as we know it, will be over in six months."

But Born needn't have worried, his job was safe. By the time QED had been given sufficient theoretical backing (and thus actually been demonstrated, QED), two new forces had been discovered. The strong and weak nuclear forces had been observed operating at nuclear level.

Scientists soon noticed a curious resemblance between the weak nuclear force and electromagnetic force. By the 1960s a mathematical theory had been evolved that described both of these forces in one set of equations. This was known as the electroweak theory. It predicted the existence of three as yet unknown subnuclear particles (W^+, W^-, and Z^0). In 1983 these were duly discovered by the particle accelerator at CERN in Geneva. Two of the four forces had now been combined, which left only three.

QED was obviously the name of the game. The physicists now set about developing a similar theory that would incorporate the strong nuclear force—which held together the protons and neutrons of the atomic nucleus.

Unfortunately the basic nuclear particles, protons and neutrons, had by now been broken down still further. At Caltech, Gell-Mann had discovered that these elementary particles in fact consisted of even more elementary particles. With characteristic erudition he named these quarks, after the cryptic quotation: "Three quarks for Muster Mark!" (This appears in James Joyce's *Finnegans Wake*—the modernist masterpiece that Gell-Mann liked to read in his spare time, which is even more difficult to understand than the universe.)

Once more the sand was beginning to spill between the fingers of the theorists who were convinced that they had it all within their grasp. Quarks required a new theory to explain how they interacted, and this was duly worked out (called QCD). The theorists then quickly set about combining QCD with electroweak theory before anything else could be discovered. A set of equations was worked out, and labeled grand unified theory (GUT). But this GUT did not in fact grandly unify everything, as might be supposed. In their haste, the theorists appeared to have forgotten all about gravity.

Hawking set about the horrendously difficult task of trying to remedy this, by coming up with a set of equations that would link gravity with the other basic forces. As he put it: "If we find the answer to that, it would be the ultimate triumph of human reason—for then we would know the mind of God."

(Knowledge of this elusive entity, and how it works, also has a long history. Pythagoras was the first to lay down the law that the mind of God *had* to conform to mathematics, back in the fifth century B.C.)

The hunt was on. But where to start? Supergravity N = 8 was ruled out as far too difficult to use, when it posited no less than 154 different types of elementary particles (less than three dozen of which have yet been discovered). It was found that even the simplest calculation, using full computer power, would take four years.

Superstring theory then took over as the number one suspect. But this too soon began spawning mind-boggling complexities. These included no less than twenty-six dimensions. (To accommodate such a seeming impossibility, each point of space must be viewed as a twenty-two-dimensional knot of space curled up and compacted so tightly that it is only apparent at under 10^{-13} (ten trillionths) of a centimeter). As if this wasn't enough, wormhole theory also emerged. According to this, black holes disappear into other universes, where they emerge as white holes spewing out all that they have gorged. (Fortunately, this theorizing well beyond the call of duty has now been reined in. White holes were a hole too far, it seems. But wormhole theory as such continues to riddle the cheese of multiple universes.)

Too little, too late. Despite vain attempts at simplification, many have given up on superstring theory as a possible theory of everything. Indeed, some scientists have begun to wonder if perhaps the entire quest may be in vain—though they have yet to attain the state of resignation reached by the philosophers. Science won't give up that easily.

Perseverance or obstinacy? According to the scientists, a TOE (or GUT) will certainly be discovered one day. There's only one snag. Short of a miracle, it will probably be so complicated that it will be incomprehensible.

(In which case, we'll be back where we started.)

But miracles do happen. In 1987 Hawking finally finished his popular book on cosmology, and it was accepted by Bantam. The book's full title was *A Brief History of Time: From the Big Bang to Black Holes,* and it was published on April Fool's Day in 1988. Bantam might not have specialized in publishing science books, but interest in cosmology was growing. They "confidently hoped" that Hawking's book would clear the hurdle into five figures.

The rest is history. From the word go, *A Brief History of Time* was a runaway success. Within ten years it was to be translated into thirty languages and sell six million copies worldwide. Why, nobody really knows. All kinds of theories have been put forward. Everybody felt they ought to know a bit about science, and this was their chance to buy (if not necessarily read) a good popular book on the subject by the best man in the business. It added intellectual street cred to one's coffee table. It made a perfect Christmas/birthday/thank-you present for grandfathers/grand-children/nephews/ uncles/the apparently il-literate generation who only seemed to be in-terested in noise and computers. It was user friendly; it was ideal for prize giving. There was a need for a new Einstein. Women gave it to men. Women read it (even if men didn't). . . . The theories abounded, and market re-searchers went into overdrive. (They wanted to discover how to do the *next* one.)

One thing everyone seemed to agree on: People bought the book, but they didn't actu-ally read it. They were too busy, too tired, had better things to do, etc. But this just isn't true. Of all the millions of copies sold, a few at least have been read from cover to cover. The impact on the (mainly young) people who made it to page 182 has been tremen-dous. It's no exaggeration to say that this book has created a new generation of scien-tists. Future Nobel prize–winners will recall:

"Then one day I read *A Brief History of Time*, and I knew what I wanted to do." This is how such a book changes the world.

So what about the book itself? For a start, it's very readable. And needless to say it's well informed. The concepts are of course difficult, and difficult to simplify without rendering them simplistic. Hawking manages this. A sample of chapter headings indicates what it's about: The Expanding Universe, Black Holes, The Origin and Future of the Universe, The Unification of Physics.

The book concludes by examining some philosophical questions—at the same time castigating philosophers "who have not been able to keep up with the advance of scientific theories." Hawking's musings may disappear down a few philosophic black holes, but they are interesting and of relevance. This is how a great modern scientist at the cutting edge of his field *thinks*. The few philosophical assumptions made by modern scientists may be wonky or just plain wrong—but they are *used,* and they are *productive*. They have produced most of the finest thinking of our time. So does philosophy matter at all to science? Hawking appears to think that it does—ultimately.

In his conclusion to *A Brief History of Time*, Hawking discusses such matters as the nature of God and unified theories (theories of everything). Whether or not these two problematic entities exist is not examined, or even considered relevant. (Hawking is a great believer in the latter, but not the former.) However, he does raise a fundamental philosophical point: "The usual approach of science of constructing a mathematical model cannot answer the question of why there should be a model for the universe to describe." (Wittgenstein, the philosopher Hawking particularly derides, in fact posed this question more succinctly over seventy years ago: "It is not *how* things are in the world that is mystical, but *that* it exists.")

Hawking asks: "Is the unified theory so

compelling that it brings about its own existence?'' Again, this is hardly a novel idea. Medieval philosophers argued that the idea of perfection must include the idea of existence, claiming this as a proof of God's existence. In Hawking's universe (or universes: an impossible but seemingly necessary plural) there is not much room for God. Though He did have a choice in creating the universe, even if this choice boiled down to no choice—because the universe *had* to be created, and had to be created in the way it was created. Why? ''There may well be only one, or a small number, of complete unified theories, such as the heterotic string theory, that are self-consistent and allow the existence of structures as complicated as human beings who can investigate the laws of the universe and ask about the nature of God.'' Such a theory is unified in the same way as a snake swallowing its tail.

*　*　*

After the publication of his best seller, Hawking quickly became a celebrity. The little man in the motorized wheelchair was pointed out as one of the sights of Cambridge. When he was there, that is. For Hawking was now hugely in demand all over the world. Trips abroad, and honors, were frequent. Jane now had a teaching job, which kept her in Cambridge during the term, so Hawking was accompanied abroad by his nurse, Elaine Mason. Jane's position had shifted. A TV film was made about Hawking entitled *Master of the Universe*. Jane saw it as her role "simply to tell him he's not God."

The outcome was perhaps inevitable. In 1990 the marriage between Jane and Stephen Hawking broke up. Hawking moved into a flat with his nurse Elaine, who was still the wife of his friend David Mason, the computer engineer.

Bitterness was inevitable. No one (meaning everyone) was to blame. It was all very scientific: The more complex the situation became, the more difficult it was to account for it. Yet there is no unified theory for human emotions. (Perhaps the theory of everything will end up being the theory of everything except what matters.)

From superstring to tinsel. In 1990 Hawking ended up in Hollywood, where he met Stephen Spielberg. They each admired the other's work. Spielberg promised to sponsor a film of *A Brief History of Time*. Hawking suggested it should be called *Back to the Future 4*. They promised to keep in touch.

Filming eventually began at Elstree Studios near London, complete with a meticulous mock-up of Hawking's office at DAMTP. Back at Cambridge, resting like any other actor, Hawking began pondering on his chances of an Oscar—"best supporting role to the universe." But sadly, the universal studios where he worked qualified him to compete only for a mere Nobel (a sort of Oscar for people who don't make it in the *real* world). Hawking was obviously quite interested in the Nobel Prize (i.e., it has the biggest entry in the index of *A Brief History of Time*). But his chances of winning it are remote indeed.

Why? As in any field of scientific endeavor, theories abound. According to one, Alfred Nobel, the Swedish dynamite tycoon who founded the prize, was cuckolded by a cosmologist. Hence, he decreed that his prize should be open to all scientists *except* cosmologists. Even so, the physics prize has a couple of times been awarded to cosmologists. But another, more explicit rule states that the scientific prizes should be given for science. In those primitive days at the turn of the twentieth century when Nobel founded his prize, science was limited to something you could *prove*. And this had to be done by observation or experiment—baffling theoretical arguments were not considered sufficient. Hawking's work cannot be proved. ("I was there, I saw the beginning of the universe.") Indeed, science is still unable even to prove the existence of black holes.

Not for nothing does Hawking work in the Department of Applied Mathematics and Theoretical Physics. If his work was proved, it could become practical and he might lose his office. It is in this office that Hawking has done much of his most profound thinking (with the sign QUIET PLEASE, THE BOSS IS ASLEEP on the door). Perhaps this is how we should best picture him. A small figure slumped in his motorized wheelchair, with its computer screen, mirror, complex wires, and clicking gadgetry, Hawking silently matches minute calculation to vast theory. On the desk facing him is another computer screen, and piles of papers. Beyond, the large poster of Marilyn Monroe gazes down tenderly on her intellectual charge. Lost to his surroundings, Hawking pits his mind against the limits of the universe. Occasionally an assistant or a nurse quietly enters, and leaves again unnoticed.

At four o'clock precisely, a daily ritual is

enacted. Tea time. Hawking is wheeled down the corridor to the common room, where portraits of previous Lucasian professors line the walls. Here lively exchanges take place among the assembled youthful researchers. The appearance of this company has been likened to a "rock group on a bad day," and their language is equally incomprehensible to normal human beings. The central figure of this group sits in his wheelchair wearing a bib. His cup is held by a nurse, who rests one hand on his forehead, lowering his head so that he can drink. His glasses slip forward down his nose, and his slack lips slurp at the tea as the young voices debate earnestly around him. Sometimes the conversation breaks off, and one of the group writes out a mathematical formula on the Formica table-top. ("When we want to save something we Xerox the table," Hawking once told a visitor.)

Occasionally the group turns to the tiny collapsed figure in the wheelchair, and he taps out a reply, which emerges in the un-nerving voice from his synthesizer. One of the group passes a typical bad-taste student comment, and the figure in the wheelchair beams his famous broad grin. He is in his element: the center of his own mathematical universe, already the stuff of legend.

Big Moments in the History of the Universe

Approx. 15 billion years ago	Big Bang
10^{-43} secs. later	Gravitational force separates as distinct entity from combined forces of the universe
10^{-36} secs.	Universe the size of a pea Temperature $10^{28}°C$
10^{-35} secs.	Electromagnetic force separates as distinct entity

10^{-12} secs.	Inflation begins
	Universe predominantly radiation
10^{-10} secs.	Weak nuclear force separates from electromagnetic force
1 sec.	Temperature falls to $10^{10}°C$
5 secs.	Formation of first nuclei
1,000 years	Predominance of matter over radiation
1,000,000 years	Formation of first atoms
1 billion years	Appearance of first galaxies
5 billion years	Appearance of Milky Way Galaxy
10 billion years	Appearance of Solar System
14,999 billion years	Appearance of hominids on Earth

15 billion years	Appearance of Stephen Hawking
20 billion years?	Universe reaches maximum expansion
35 billion years?	Increasingly rapid proliferation of singularities (black holes)
40 billion years?	The Big Crunch

Suggestions for Further Reading

Hawking, Stephen. *A Brief History of Time* (Bantam, 1990)—The global best seller in which the man himself explains the universe to the world.

Hawking, Stephen (ed.). *A Brief History of Time: A Reader's Companion* (Bantam, 1992)—The Hawking story in the words of friends, family, and the man himself.

White, Michael and John Gribben. *Stephen Hawking: A Life in Science* (Plume, 1993)—The nearest thing to a full-scale biography.

Kraus, Gerard. *Has Hawking Erred? An Appraisal of "A Brief History of Time"* (Janus, 1994)—The opposition point of view.

Hawking, Stephen. *Black Holes and Baby Universes and Other Essays* (Bantam, 1994)— The latest news on time, the universe, and all that.

NEWTON AND GRAVITY

A good case can be made for Isaac Newton being the finest mind humanity has yet produced. His theory of gravity offered his contemporaries their first glimpses of how the universe actually works, and his mathematics enabled later generations to walk on the moon. Today, we know that gravity is what keeps our feet on the ground, but how many of us know how Newton's greatest discovery really works? In *Newton and Gravity* Paul Strathern encapsulates several of Newton's more world-altering discoveries, explaining in lively prose their cultural context as well as Newton's early obsession with science (bordering on dementia) that made his revolutionary vision possible. Just a few of the big ideas covered here are:

- Newton's discovery of calculus at age twenty-three
- Why one of the greatest human insights of all time was in fact a hunch, and how it actually works
- Why it took Newton twenty years after his discovery

to reveal to the world the secret of gravity and plane-
tary motion

Ideal for the intelligent reader eager to understand
better how and why the universe works the way it
does, *Newton and Gravity* is a fascinating refresher
course that makes physics not only fun but shockingly
easy to understand.